California
Science

Science Content Support
Grade 1

Harcourt
SCHOOL PUBLISHERS

Visit *The Learning Site!*
www.harcourtschool.com

Printed in the United States of America

ISBN-13: 978-0-15-352280-2
ISBN-10: 0-15-352280-1

2 3 4 5 6 7 8 9 10 022 17 16 15 14 13 12 11 10 09 08

Contents

Ready, Set, Science!

Unit 1 • States of Matter

Unit 2 • Plants and Animals Meet Their Needs

Unit 3 • Weather

Vocabulary

Welcome to *Science!*

Find these things in your book.

1. What animal is on the cover of your book?

_____

_____

_____

2. What bridge is in the picture on page 52?

_____

_____

_____

3. Look in the Health Handbook. It is in the
back of your book. On what page do you
find a skeleton?

_____

_____

_____

4. What is the first word in the glossary?

_____

_____

_____

Name_____

5. What is the last word listed in the index?

_____

- -

_____

6. Find an activity that you would like to try. Write its title and page number.

_____

- -

_____

7. Where was the postcard on page 51 sent from?

_____

- -

_____

8. What is the first vocabulary word on page 133?

_____

- -

_____

9. Find a picture of something you like. Tell about it.

_____

- -

_____

Classify Words

To better understand a word, you can think about words that tell more about it.

The words **inch**, **foot**, and **mile** all tell about **measurement**.

Read the words in the box. Write the words that can tell about investigation skills on the lines.

observe	**compare**	**soil**
plant	**predict**	**sun**
tell		

investigation skills

PLANT _____ _____	_____ _____
_____ _____	_____ _____

Take Notes

When you read, write down important words and sentences. This is called taking notes. A learning log can help you record, organize, and remember your notes.

As you read about investigation skills, write one more sentence about this topic under take notes. Write your own thoughts about this new sentence under make notes.

take notes	make notes
• **Scientists use investigation skills to find out information.**	• **Scientists learn new things all the time.**
• **People use investigation skills, too.**	• **I can be like a scientist.**
• _____	• _____

Name_____

Lesson 1—How Do We Use Investigation Skills?

1. **Investigation Skill Practice–Observe**

Observe an object in your classroom. Draw it. Tell about it.

This is a _____. It is _____ and _____.

2. (Focus Skill) **Reading Skill Practice–Main Idea and Details**

Fill in the chart below.

Investigation Skills

Main Idea
Investigation skills help people find out information.

detail
Scientists use investigation skills when they do **A** _____.

detail
You can **B** _____, or use your senses to find out about things.

© Harcourt

Name_____

Science Concepts

3. Label each picture. Use a word from the word box.

observe	classify	sequence	measure

_____

- - - - - - - - - - - - - - - - - -

| 1 | 2 | 3 |

_____

- - - - - - - - - - - - - - - - - -

_____

- - - - - - - - - - - - - - - - - -

_____

- - - - - - - - - - - - - - - - - -

4. Observe your teacher. Draw a picture to tell about him or her.

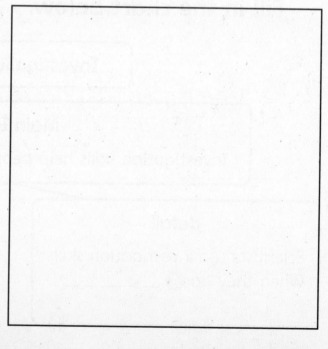

© Harcourt

Name _____

Date _____

Writing Sentences

You can use these words together in a sentence.

forceps	magnifying box

We used **forceps** to separate the rocks and to place them in the **magnifying box**.

Write a sentence that includes the words below. Use what you know about the meanings of the words to help you.

1. hand lens, science tools

_____

_____

2. measuring cup, thermometer

_____

_____

3. tape measure, ruler

_____

_____

© Harcourt

Name_____

Date _____

Build Vocabulary

You can use a word web to help you understand and remember new words. A word web has one word in the center. The words or ideas around the center tell more about it.

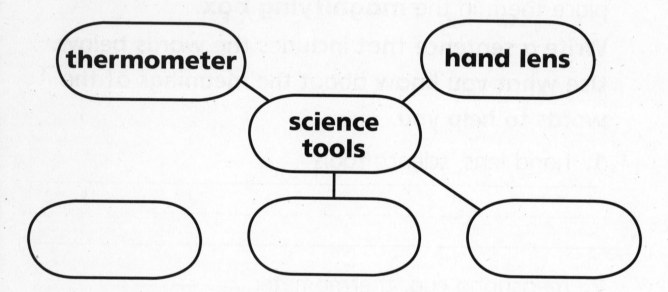

Use the clues to add three more words about science tools to the word web.

- You can use a _____ to measure things.
- You can separate things with _____.
- You can use a _____ to place drops of liquid.

© Harcourt

Name _____

Date _____

Lesson Quick Study

Ready, Set, Science!, Lesson 2

Lesson 2—How Do We Use Science Tools?

1. **Investigation Skill Practice–Compare**

Choose two objects from your pencil box. Draw them. Compare them. Tell about each object.

2. **Focus Skill** **Reading Skill Practice–Main Idea and Details**

Fill in the chart below.

Main Idea
You can use science tools.

detail
Science tools help you do investigations. They help you observe, compare, and **A** _____ things.

detail
Use a **B** _____ or magnifying box to see very small things.

© Harcourt

Use with Ready, Set, Science! (page 1 of 2) **Science Content Support** **CS 7**

Name_____

Science Concepts

3. Circle the tool you would use to do each job.

You have a block and a ball. You want to know which one has more mass. You should use a _____.

ruler   balance

You want to put three drops of water in a cup. You should use a _____.

measuring cup   dropper

You have a leaf. You want to find out how long it is. You should use a _____.

ruler   measuring cup

4. Which tool would you use to measure the size of a ball? Circle the correct picture.

   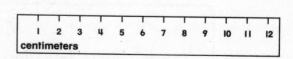

Why would you use this tool?

_____

- -

_____

Word Meanings

You can use these words to show where
objects are.

above	below	over

The roof is **above** my head.

1. Read each sentence below. Draw a picture that
   shows its meaning.

The circle is **above** the square.	The triangle is **below** the circle.	A bird flies **over** a pond.

2. Circle the words that tell where things are.

above   smooth   red   large

left   bumpy   below   right

Use Visuals

Photographs, drawings, and pictures are called visuals. Visuals can tell you more about what you read. When you look at a visual, search for details in the picture. Asking questions about these details will help you learn even more information.

Look at the visual on page 24. Answer the questions below.

1. Where is this picture taken from? How do you know?

_____

- -

_____

_____

- -

_____

2. Pretend you are in a helicopter, flying above your home. How does your house look? Write a sentence describing what you would see.

_____

- -

_____

_____

- -

_____

© Harcourt

Lesson 3—How Can We Describe the Position of Objects?

1. Investigation Skill Practice–Tell

Draw a toy in each box. Tell where the toys are.

The _____ is next to the _____.

2. Focus Skill Reading Skill Practice–Main Idea and Details

Fill in the chart below.

Main Idea
You can use words to tell where things are.

detail
Ⓐ _____ means that a thing is at a higher place.

detail
Ⓑ _____ means that a thing is at a lower place.

Name_____

Science Concepts

3. Read the directions to complete the picture.

Draw a ball next to the sandcastle.

Draw a kite below the cloud.

Draw a plane over the cloud.

4. Answer each question by looking around you.

Name something behind you. _____

Name something next to you. _____

Name something above you. _____

© Harcourt

Position of Objects

Use the pictures to answer the questions below.

1. Which object is to the left of the sun and below the flower?

_____

- -

_____

2. Which object is above the sun and to the left of the fish?

_____

- -

_____

3. Circle the object that is below the sun and to the right of the cat.

4. Underline the object that is above the sun.

Position of Objects

Draw the objects below in the correct places.

1. Draw a circle below the star and to the right of the letter A.

2. Draw a square above the star and to the left of the letter B.

3. Draw a triangle to the right of the star and above the letter C.

4. Draw a heart to the left of the star and below the letter D.

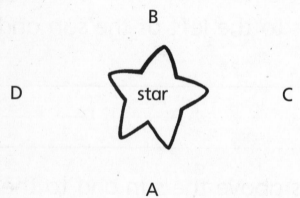

Use the pictures above to answer the question.

5. Which object is below the star and to the left of the letter A?

_____

- -

_____

Classify Words

You can classify words by what they mean. For example, you can classify **observe**, **predict**, **ruler**, and **measuring cup** into two groups:

investigation skills	science tools
observe	ruler
predict	measuring cup

Look at the words below. Use the chart to classify them.

question	words	test
plan	bar graph	draw pictures

investigation steps	ways to record observations
_____	_____
_____	_____
_____	_____
_____	_____
_____	_____
_____	_____
_____	_____
_____	_____

© Harcourt

Anticipation Guide

Using an anticipation guide will help you get ready to read. An anticipation guide is a list of statements about a topic. Reading the list of statements helps you to think about the topic and learn more.

Read the list of statements below. Circle T if you think the statement is true. Circle F if you think it is not correct.

T	F	Scientists follow steps to test things they want to learn.
T	F	Scientists do not think and plan before they test.
T	F	Scientists record what they observe.
T	F	Scientists only use pictures to record what they observe.
T	F	You can use a bar graph to record what you observe.
T	F	A bar graph cannot compare things.

Use this anticipation guide as you read about scientists. Change any answers you did not anticipate correctly.

© Harcourt

Lesson 4—How Do Scientists Work?

1. Investigation Skill Practice–Plan an Investigation

You want to know if a rock or a leaf will float. Which plan could you follow? Circle it.

2. ⭐ Reading Skill Practice–Main Idea and Details

Fill in the details about doing a test.

_____

_____. Then ask a question.

Form a _____.

Plan a fair _____.

_____ the test.

Draw _____. Tell what you think.

© Harcourt

Name_____

Science Concepts

3. You have a feather and a block. You want to know which object will hit the ground first. How can you test your idea?

_____

- -

_____

4. Can a penny and a feather be moved by air? Number the steps below from 1 to 5 to show how you could test this idea.

_____ Write a plan to test your idea.

_____ Ask the question: **Can air move a penny and a feather?**

_____ Record what you observe.

_____ Share your results with a classmate.

_____ Follow your plan.

© Harcourt

Bar Graph Practice

Use the data from the tally chart. Fill in the bar graph below. Give the bar graph a title. Label the axes.

Favorite Kinds of Fruit	
bananas	~~HHH~~
apples	~~HHH~~ IIII
grapes	III
strawberries	~~HHH~~ II

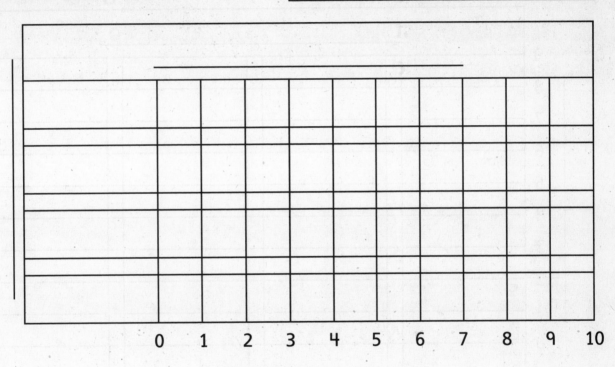

0 1 2 3 4 5 6 7 8 9 10

_____

Answer the question below.

1. How many kids like grapes and bananas?

_____

- -

_____

© Harcourt

Name_____

Date _____

Bar Graph Practice

Use the data from the chart. Fill in the bar graph below. Give the bar graph a title. Label the axes.

Favorite Kinds of Sports	
baseball	8
football	6
basketball	4
soccer	3

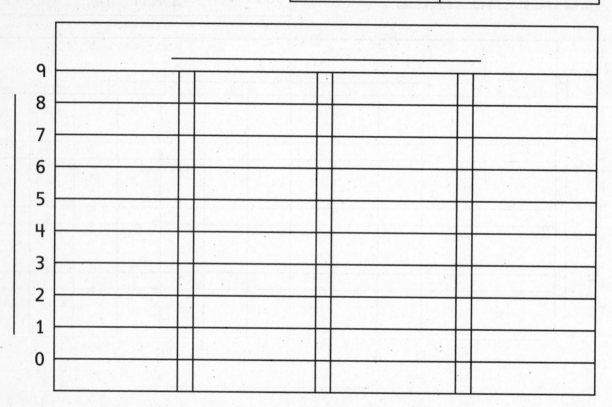

Answer the question below.

1. How many kids like football and soccer?

_____

- -

_____

© Harcourt

Context Clues

Choose a word from the box to finish each sentence. Write the word on the line. Use the other words in the sentence to help you.

matter	property	mass	
solid	liquid	gas	texture

1. Toys, air, and water are kinds of _____.

2. An elephant has a lot of _____, but a pencil only has a little.

3. Three forms of matter are _____, _____ _____, and _____.

4. Tree bark can have a very rough _____.

5. One _____ of water is that it feels wet.

Name_____

Date _____

Build Vocabulary

You can use a word web to help you
understand new words. A word web has
a word in the center. The words or ideas
around it link to the word in the center.

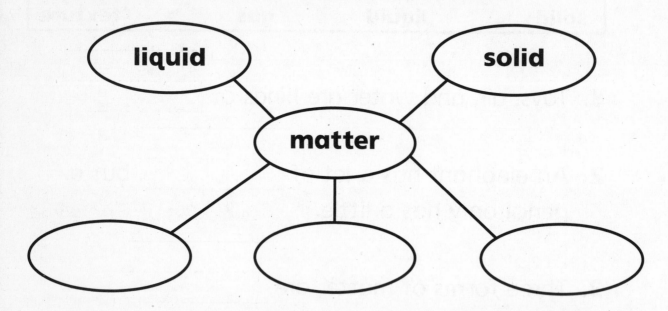

Use the clues to add more words about matter.

- Matter comes in different forms. It can be a
  solid, a liquid, or a _____.

- All matter has two main properties. Matter takes
  up space and has _____.

- Toys, air, and water are all matter. Name another
  kind of matter.

© Harcourt

Lesson 1—What Is Matter?

1. Investigation Skill Practice–Classify

Write about ways to classify these objects.

I could classify the objects by
_____

- -

_____ .

2. ⭐(Focus Skill) Reading Skill Practice–Compare and Contrast

Fill in the chart below.

alike	different
A _____ is matter.	Matter can be different forms, such as a solid, a liquid, or a **B** _____.
	Matter can be different sizes, such as **C** _____ and **D** _____.

Name _____

Science Concepts

3. Label each picture. Use a word from the word box.

| solid | liquid | gas |

_____ _____ _____

- - - - - - - - - - - -   - - - - - - - - - - - -   - - - - - - - - - - - -

_____ _____ _____

4. Draw two different kinds of matter you see in the classroom. Write about their properties.

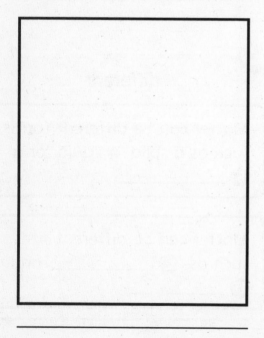 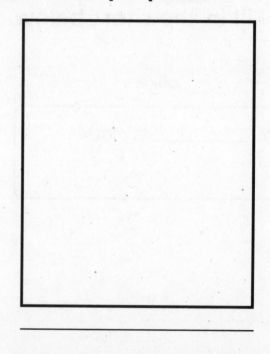

_____ _____

- - - - - - - - - - - -   - - - - - - - - - - - -

_____ _____

© Harcourt

Classify Words

You can classify words by what they mean. For example, you can classify **ruler**, **observe**, **infer**, and **balance** into two groups:

investigation skills	science tools
observe	ruler
infer	balance

Look at the words below. Use the chart to classify them.

solid	length	liquid
gas	weight	mass

forms of matter	measurements
_____	_____
_____	_____
_____	_____

© Harcourt

Name_____

Date _____

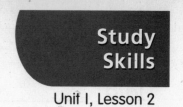

Use a K-W-L Chart

New ideas are easier to understand when you write what you learn. A K-W-L chart can help you record ideas before and after you read.

- Use the K column to write what you know about solids.

- Use the W column to write what you want to know about solids.

- Use the L column to write what you learned about solids.

K what I know about solids	W what I want to know about solids	L what I learned about solids
A solid is a kind of matter.	How big is a solid?	I can measure a solid.
My desk is a solid.		

© Harcourt

Lesson 2—What Can We Observe About Solids?

1. Investigation Skill Practice—Compare

Look at the solid objects. Then answer the question.

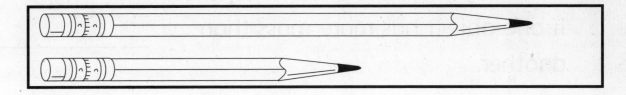

How are the objects the same and different?

_____

- -

_____

2. (Focus Skill) Reading Skill Practice—Main Idea and Details

Fill in the chart below.

> **Main Idea**
> A solid is matter that keeps its shape.

> **detail**
> You can **A** _____ and feel a solid.

> **detail**
> You can **B** _____ solids with a ruler, a balance, or a scale.

Name_____

Science Concepts

3. Which tool would each person use to do their job?

ruler	balance	scale

Alex is a builder. He wants to measure the length of a window.

_____

Lisa is a scientist. She wants to know if one object has more mass than another.

_____

Sara works at a market. She wants to know how much a bag of apples weighs.

_____

4. Draw a solid you see in your classroom. How could you measure it?

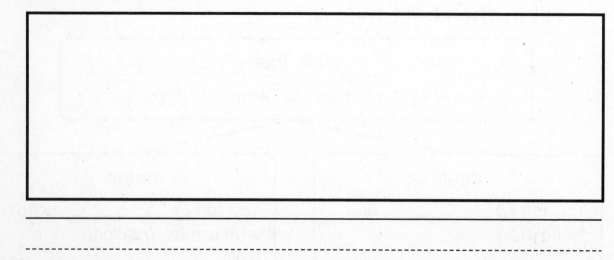

_____

_____

© Harcourt

Naming and Acting Parts of Sentences

Every sentence has a part that names
what the sentence is about.

    Leaves float on water. This sentence is
    about **leaves**.

Every sentence has a part that tells what is
happening.

    Leaves **float** on water. This sentence tells the
    reader that leaves can **float**.

**Draw a line from the naming part to
the action part to form a sentence
that makes sense.**

The new boat   sink to the bottom.

The cup   measures the liquid.

A liquid   floats across the lake.

The rocks   takes the shape of
                     its container.

© Harcourt

Name_____

Date _____

Preview and Question

Asking questions as you read helps you understand new information.

Use the chart below to write two more questions about liquids. Look for the answers to the questions as you read the lesson.

liquids	
questions	**answers**
What is a liquid?	Liquid is matter that flows.
What shape is a liquid?	

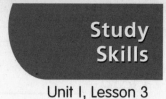

© Harcourt

Lesson 3—What Can We Observe About Liquids?

1. Investigation Skill Practice–Predict

Predict which objects will float. Circle them. Test your prediction.

2. Reading Skill Practice–Main Idea and Details

Fill in the chart below.

> **Main Idea**
>
> A liquid is matter that **Ⓐ** _____.
> It **Ⓑ** _____ its own shape.

> **detail**
>
> You can
> **Ⓒ** _____ and
> feel a liquid.

> **detail**
>
> Some solids float in
> liquids, and some solids
> **Ⓓ** _____ in
> liquids.

© Harcourt

Name_____

Science Concepts

3. Look at the fish tank.

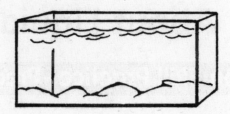

What would happen if you put a toy boat in the water? Draw a picture to show your answer.

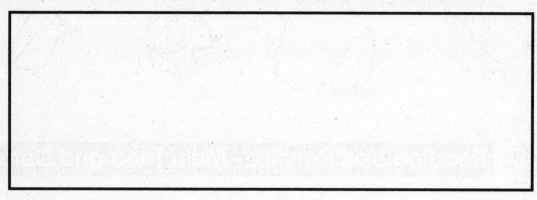

4. Circle the liquid.

air   **cup**   **milk**

How do you know it is a liquid?

_____

- -

_____

Name one more liquid.

_____

- -

_____

Classify Words

You can list words that tell more about something.

Juice, **water**, and **flows** tell more about **liquids**.

Solid, **liquid**, and **gas** tell more about **matter**.

Look at each word below. Write the word from the box that tells more about it. Then write another word of your own.

gas	shape	measure

1. air _____   _____

2. property _____   _____

3. balance _____   _____

Name_____

Date _____

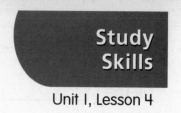

Take Notes

Take notes when you read. Write down important words and sentences that you want to remember. A learning log is a chart you can use to record your notes.

Read the lesson about gas starting on page 88. Write one more important sentence under take notes. Write your own thoughts under make notes.

take notes	make notes
• Gas does not have its own shape.	• A liquid does not have its own shape either.
• Gas spreads out to fill a container.	• _____
• _____	• _____

© Harcourt

Lesson 4—What Can We Observe About Gases?

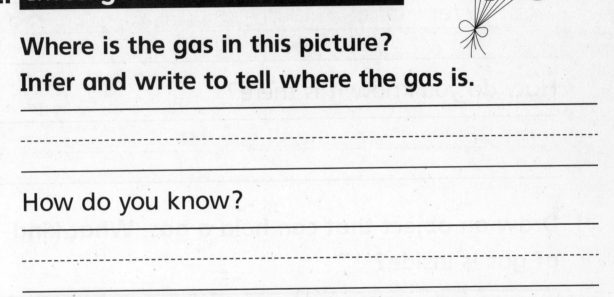

1. Investigation Skill Practice–Infer

Where is the gas in this picture?
Infer and write to tell where the gas is.

_____

- -

_____

How do you know?

_____

- -

_____

2. Reading Skill Practice–Main Idea and Details

(Focus Skill)

Fill in the chart below.

> **Gases**

> **Main Idea**
> A gas is matter that **A** _____ its own shape.
> It **B** _____ to fill its container.

> **detail**
> Air is made of **C** _____.
> Often you can not **D** _____,
> smell, or feel air.

> **detail**
> You can see and feel what moving
> air does.

© Harcourt

Name_____

3. What is the gas in this picture?

_____

- - - - - - - - - - - - - - - - - - -

_____

How do you know it is there?

_____

- - - - - - - - - - - - - - - - - - -

_____

4. Draw an object that can hold a gas. What kind of gas is inside?

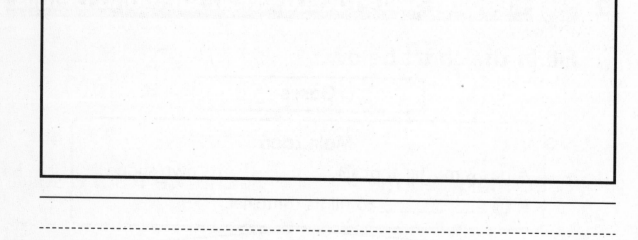

_____

- - - - - - - - - - - - - - - - - - -

_____

© Harcourt

Words that Mean the Same

Record is another way to say write down.

Match each clue to a word in the box. Write the word in the puzzle.

mixture	evaporation	heat
cool	freeze	melt

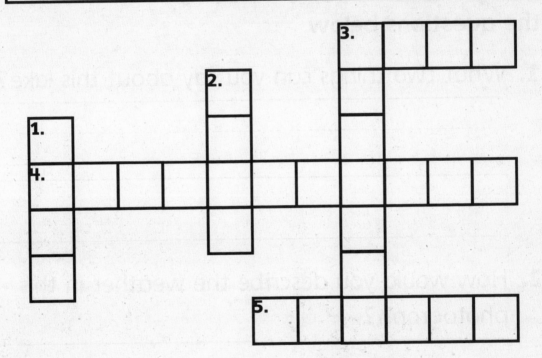

Down

1. to make warmer

2. to make colder

3. mix of two or more kinds of matter

Across

3. change from a solid to a liquid

4. change from a liquid to a gas

5. change from a liquid to a solid

Name_____

Date _____

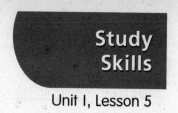
Use Visuals

Photographs, drawings, and pictures are called visuals. Visuals can tell you more about what you read. Asking questions about visuals will help you learn more information.

Look at the photograph on page 96. Answer the questions below.

1. What two things can you say about this lake?

_____

- -

_____

_____

- -

_____

2. How would you describe the weather in this photograph?

_____

- -

_____

3. What change do you think happened to make the lake look like this?

_____

- -

_____

© Harcourt

Lesson 5—How Can Matter Change?

1. Investigation Skill Practice–Tell

You place an ice cube outside on a sunny day.
Tell what will happen.

_____

- -

_____

Why did this happen?

_____

- -

_____

2. (Focus Skill) Reading Skill Practice–Cause and Effect

Complete the chart to show cause and effect.

Changes in Matter

cause   effect

cause	effect
Water is cooled.	The water changes to **A** _____.
Ice is placed in the hot sun.	The ice **B** _____.
Water becomes very **C** _____.	The water changes from a **D** _____ to a gas.

Name_____

Science Concepts

3. Circle the picture that shows what happens next.

Sam left his ice cream on the table and went out to play. What happens next?

4. Use the words in the word box to label what is described in the chart.

melt	freeze	mixture
_____	_____	_____
a mix of nuts, raisins, and cereal	a stream during winter	a piece of ice in a hot pan

© Harcourt

Context Clues

Context Clues can help you understand the meaning of a new word.

Water **freezes**, or **changes from a liquid to a solid**, when you cool it.

The words **changes from a liquid to a solid** help you understand what the word **freezes** means.

Read each sentence and its answer choices. Write the correct word on the line.

1. The solids left over after matter burns

are _____.

ashes   **smoke**

2. Matter changes to new matter when

it is _____.

cooked   **fire**

3. Matter changes to new matter when

it is _____.

cooled   **burned**

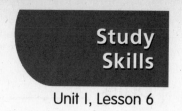
Anticipation Guide

Using an anticipation guide will help you get ready to read. An anticipation guide is a list of statements about a topic.

Read the list of statements about changes to matter. Circle T if you think the statement is true. Circle F if you think the statement is not correct.

T	F	Fire changes matter into new matter.
T	F	You can change ashes and smoke back to wood.
T	F	Cooking can change matter.
T	F	When you toast bread, the color changes.
T	F	When you cook eggs, their shape, color, and texture stay the same.
T	F	Cooked eggs can never go back to uncooked eggs.

Use this anticipation guide as you read the lesson. Find out if your thoughts about matter change.

Lesson 6—What Are Other Changes to Matter?

1. Investigation Skill Practice–Observe

Write a sentence describing the changes in matter you would observe.

A leaf falls in a campfire.

_____

- -

_____

You make a grilled cheese sandwich.

_____

- -

_____

2. (Focus Skill) Reading Skill Practice–Cause and Effect

Write the cause of each effect.

cause	effect
_____	The wood changes into ashes and smoke.
_____	An egg changes color, size, and shape.

Name_____

Science Concepts

3. Circle the examples that show a change in matter.

A piece of paper burns.

A piece of bread is toasted.

A flag blows in the wind.

4. Tell how each object changes.

A hamburger patty is cooked.

_____

- -

_____

A log is burned.

_____

- -

_____

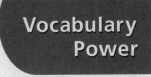
Rhyming Words

Rhyming words have the same ending sounds.

Sun rhymes with f**un**, b**un**, sp**un**, and r**un**.

Tr**ee** rhymes with fr**ee**, b**ee**, kn**ee**, and s**ee**.

Look at each word. Write two words that rhyme.

1. roots _____

_____

2. leaves _____

_____

3. plants _____

_____

4. need _____

Look at each set of words you wrote. Circle the letters that stand for the same ending sound. Remember that different vowel patterns can have the same sound.

© Harcourt

Name_____

Date _____

Use Visuals

Visuals can tell you more about what you read. A visual can be a drawing or a photograph. Asking questions about visuals will help you learn more information.

Look at the photograph on page 137. Answer the questions below.

1. What is the girl doing in the picture?

_____

- -

_____

2. What do the plants in this picture look like?

_____

- -

_____

3. Do you think the plants in this picture will grow? Tell why or why not.

_____

- -

_____

_____

- -

_____

© Harcourt

Name _____

Date _____

Lesson 1—What Do Plants Need?

1. **Investigation Skill Practice–Predict**

**Look at the pictures. Predict what will
happen to each plant after a few days.**

I predict this plant will   I predict this plant will

_____   _____

- - - - - - - - - - -   - - - - - - - - - - -

_____ .   _____ .

How could you check your prediction?
_____
- -
_____

2. (Focus Skill) **Reading Skill Practice–Main Idea and Details**

Circle the things plants need to live and grow.

light   water   animals   air

Name_____

3. People water houseplants. Most plants grow outside. Write labels that tell how the plant in this picture gets water.

4. Draw a picture of a plant. Label the **leaves** and **roots**.

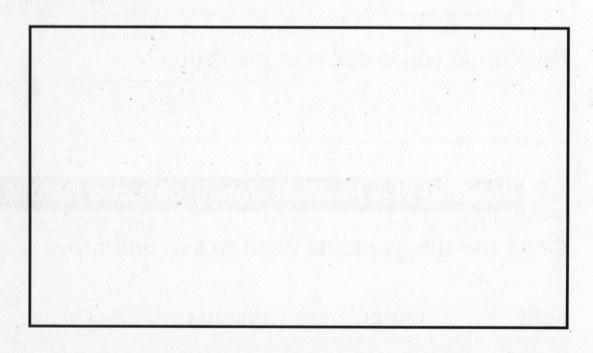

© Harcourt

Name_____

Date _____

Naming and Doing Words

Shelter, **water**, and **drink** are words for things. Look how they are used here.

The fox lives in a **shelter**.

A giraffe stands by the **water**.

The cat takes a **drink**.

Shelter, **water**, and **drink** can also be words for what something **does**. Find out how. Complete the sentences with words from the box.

shelter	water	drink

1. Dad will _____ the plants.

2. Elephants _____ from the pond.

3. Our house can _____ us from the rain.

© Harcourt

Use a K-W-L Chart

Before you read about something new, think about what you already know about that topic. A K-W-L chart will help you organize new ideas before and after you read.

- Use the K column to write what you know about what animals need.
- Use the W column to write what you want to know about what animals need.
- Use the L column to write what you learned about what animals need.

K what I know about what animals need	W what I want to know about what animals need	L what I learned about what animals need
Animals need food.	What kind of food do animals eat?	

© Harcourt

Lesson 2—What Do Animals Need?

1. **Investigation Skill Practice–Observe**

Look at the animals. What need is each meeting? Write about what you observe.

_____

- - - - - - - - - - - - -

_____

_____

_____

- - - - - - - - - - - - -

_____

_____

- - - - - - - - - - - - -

_____

2. (Focus Skill) **Reading Skill Practice–Main Idea and Details**

List four things animals need to live and grow.

Animal Needs

_____   _____

- - - - - - - - -   - - - - - - - - -

_____   _____

_____   _____

- - - - - - - - -   - - - - - - - - -

_____   _____

Name _____

Science Concepts

3. Draw a picture of an animal in its shelter.

[ ]

4. Look at the animal's teeth. Does it eat plants or meat? How can you tell?

This animal eats _____ .

I know this because it has _____ teeth.

© Harcourt

Word Meaning

To better understand a word, you can think about words that tell about it.

The words **water**, **fish**, **plants**, and **insects** tell about a **pond**.

Add words to each list below. Then draw each environment. Include all of the words from your list in the picture.

desert	ocean	forest
dry	fish	trees
_____	_____	_____
_____	_____	_____
_____	_____	_____
_____	_____	_____
_____	_____	_____
_____	_____	_____
_____	_____	_____

Preview and Question

Asking questions as you read helps you understand new information. Use the chart below to write two more questions about where plants and animals live. Look for the answers to all the questions as you read the lesson.

Where Do Plants and Animals Live?	
questions	**answers**
Do plants and animals change because of where they live?	

© Harcourt

Lesson 3—Where Do Plants and Animals Live?

1. **Investigation Skill Practice–Observe**

Look at this desert plant.
What helps this plant live in
the desert? Write about what
you observe.

_____

- -

_____

2. (Focus Skill) **Reading Skill Practice–Main Idea and Details**

Name living things that live in each
environment.

forest	desert	ocean

Name_____

Science Concepts

3. Look at each picture. Where does the plant or animal live? How do you know?

I think this animal lives in the

_____

_ _ _ _ _ _ _ _ _ _ _ _ _ _ _ _ , because

_____

_ _

_____ .

I think this plant lives in the

_____

_ _ _ _ _ _ _ _ _ _ _ _ _ _ _ _ , because

_____

_ _

_____ .

4. Read each question. Circle the answer.

What part helps a 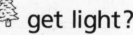 get light?

    roots   leaves

What part helps a 🐢 swim to find food?

    flippers   hard shell

Classify Words

You can classify words by what they mean. For example, you can classify **root**, **skin**, **leaf**, and **gills** into two groups:

plant parts	animal parts
root	skin
leaf	gills

Look at the words below. Use the chart to classify them.

pollen	food chain	teeth
stem	wings	environment

plant word	animal word
_____	_____
_____	_____
_____	_____

both
_____
_____
_____

© Harcourt

Connect Ideas

Connecting ideas, facts, and information help us to learn. One idea or fact can lead to the next idea or fact. You can use a flow chart to show the connection.

Read about food chains. Use sentences and pictures to complete the flow chart.

The grass uses sunlight to make its food. The grasshopper eats the grass.	The frog eats the grasshopper.

© Harcourt

Lesson 4—How Do Living Things Help Each Other?

1. Investigation Skill Practice–Observe

Look at the picture. How is the animal helping the plant? Write about what you observe.

_____

- -

_____

2. **Focus Skill** Reading Skill Practice–Main Idea and Details

Fill in the chart below.

Main Idea
Animals use plants and other animals to meet their needs.

detail
Animals use plants for a
Ⓐ _____, or to stay safe.

detail
Some animals live near other animals to stay
Ⓑ _____.

detail
Animals eat
Ⓒ _____
or other
Ⓓ _____.

© Harcourt

Name_____

Science Concepts

3. These animals are using plants and other animals to meet their needs. Circle the words that tell what they are doing.

spreading seeds

making a nest

making shelter

getting food

4. Number each animal from 1 to 4 to show its order in the food chain. Number 4 should be at the top of the food chain.

_____ hawk

_____ insect

_____ frog

_____ snake

© Harcourt

Nouns and Actions

Look at the words below. All of them have to do
with weather. Some are words for **nouns**. Others
are words for **actions**. Some words, like **snow** and
rain, can be words that name a thing or tell an
action.

nouns		actions	
sun	snow	rain	blow
rain	wind	snow	predict

The plants need
more **rain** this week.

It will **rain** tomorrow.

Draw a picture to go with each word below.

Then circle the nouns. Underline the actions.

weather	observe
cloud	predict

Name_____

Date _____

Use a K-W-L Chart

New ideas are easier to understand when you write what you learn. A K-W-L chart can help you record important ideas before and after you read.

- Use the K column to write what you know about weather.
- Use the W column to write what you want to know about weather.
- Use the L column to write what you learned about weather.

K what I know about weather	W what I want to know about weather	L what I learned about weather

Use with Unit 3.

© Harcourt

Name_____

Date _____

Lesson 1—How Does Weather Change from Day to Day?

1. Investigation Skill Practice–Predict

Look at the weather chart. Predict what the weather will be on Friday. Draw a picture.

| Monday | Tuesday | Wednesday | Thursday | Friday |

2. Reading Skill Practice–Compare and Contrast

Fill in the chart below.

alike   different

| All weather is what the **A** _____ outside is like. | Weather can be warm or **B** _____. |
| You can see and **C** _____ all weather. | Weather can also be sunny, cloudy, snowy, windy, and **D** _____. |

Name_____

3. Circle the word that describes the weather in each picture.

sunny	rainy	rainy
windy	snowy	sunny
rainy	sunny	snowy

4. Look at the weather chart. Draw pictures to show the missing weather.

Monday	Tuesday	Wednesday	Thursday	Friday
	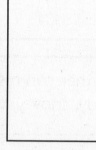			
sunny	windy	cloudy	rainy	snowy

© Harcourt

Name_____

Date _____

Multiple Meaning Words

Some words can have more than one meaning.

Do not **fall** on the ice!

The days grow shorter in the **fall**.

Each word in bold has another meaning. Use what you know about seasons to help you.

1. The toy had a **spring** that made it bounce. What else can the word **spring** mean? Draw and write a sentence about your idea.

_____

- -

_____

2. Mom uses salt to **season** her food. What else can the word **season** mean? Draw and write a sentence about your idea.

_____

- -

_____

Build Vocabulary

You can use a word web to help you understand new words and ideas. A word web has a word in the center. Around it are the words or ideas that link to the word in the center.

Add four more words to the web that tell more about winter.

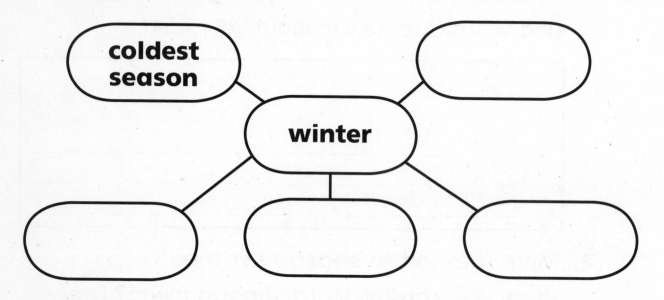

What other new words did you learn? Make a web for the seasons of spring, summer, and fall.

© Harcourt

Lesson 2—How Does Weather Change with Each Season?

1. Investigation Skill Practice–Draw a Conclusion

What season is it?

How can you tell?

_____

I think the season is _____, because

_____

_____

_____.

2. (Focus Skill) Reading Skill Practice–Sequence

Fill in the chart below.

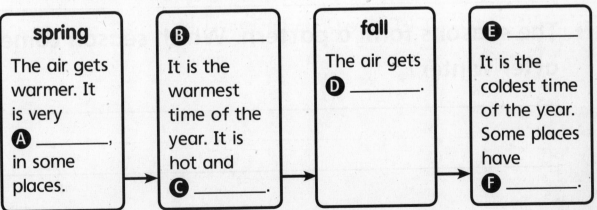

spring
The air gets warmer. It is very **A** _____, in some places.

B _____
It is the warmest time of the year. It is hot and **C** _____.

fall
The air gets **D** _____.

E _____
It is the coldest time of the year. Some places have **F** _____.

© Harcourt

Name_____

3. Think about the weather in each season. Draw and label one thing you would wear.

spring	summer	fall	winter

_____   _____   _____   _____

4. The seasons form a pattern. Which season comes after winter?

_____

_____

Name _____

Date _____

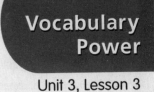

Writing Sentences

You can use these words together in a sentence.

measure   air

You can **measure** how hot or cold the **air** is.

Write a sentence that includes the words below. Use what you know about the meanings of the words to help you.

1. temperature, thermometer

_____

- -

_____

2. direction, wind vane

_____

- -

_____

3. rain gauge, rain

_____

- -

_____

© Harcourt

Take Notes

Take notes when you read. Write down important words and sentences that you want to remember. A learning log is a chart you can use to record your notes.

Read the lesson about weather starting on page 226. Write two more important sentences under take notes. Write your own thoughts under make notes.

take notes	make notes
• **Temperature is the measure of how hot or cold something is.**	•
•	•
•	•

Lesson 3—How Can We Measure Weather?

1. Investigation Skill Practice—Measure

Write the temperature shown on each thermometer.

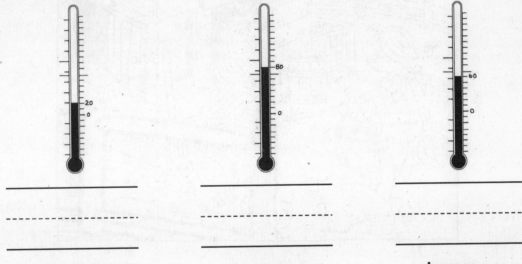

_____   _____   _____

- - - - - - - - - - - -   - - - - - - - - - - - -   - - - - - - - - - - - -

_____   _____   _____

degrees Fahrenheit   degrees Fahrenheit   degrees Fahrenheit

2. (Focus Skill) Reading Skill Practice—Main Idea and Details

Match each tool to the words that describe it.

wind vane •

rain gauge •

thermometer •

• measures how much rain has fallen

• measures temperature

• tells what direction the wind is coming from

Name_____

Science Concepts

3. Circle all the tools in the picture that measure weather.

4. You want to measure how much rain falls in a week. If you leave your rain gauge out all week, it may spill over. What can you do? Write a plan.

_____

- -

_____

- -

_____

© Harcourt

Opposites

Look at these sentences. Each sentence includes two words that are opposites. The opposites tell about the word in bold.

Water can be <u>cold</u> or <u>hot</u>.

Rain can be described as <u>light</u> or <u>heavy</u>.

Now look at the words below. Write opposites to complete the sentences. Use the words in the box.

| up | hot | dry | down | wet | cold |

1. In the **water cycle**, water vapor moves

_____ to the sky and then falls back

_____ to Earth.

2. **Land** without water will be _____, but

rain will make the land _____.

3. Desert **air** can be _____ during the day

when the sun shines and _____ at night.

© Harcourt

Connect Ideas

When you read, you learn about many
new ideas, facts, and events. One fact
or event can sometimes lead to the next
fact or event. You can use a flow chart
to show how these facts and events are
connected.

**Read about the water cycle. Use sentences to
complete the flow chart.**

| The sun makes water warm. Water evaporates. | → | Water vapor meets cool air and condenses into tiny drops. The drops form clouds. |

The water cycle begins again.

Name ______________________________

Date ______________________________

Lesson Quick Study

Unit 3, Lesson 4

Lesson 4—How Does the Sun Cause Weather?

1. Investigation Skill Practice–Plan an Investigation

You have two pieces of paper—black and white. How can you find out which paper the sun warms faster? Plan an investigation.

____

____

____

2. Focus Skill Reading Skill Practice–Cause and Effect

Fill in the chart below.

The Sun and Weather

cause		effect
The sun warms the (A) ________.	→	The (B) ________ above the land gets warmer.
The air gets warmer.	→	Warming the air causes (C) ________.
The wind moves the (D) ________.	→	Some clouds bring (E) ________.

Science Concepts

3. Draw arrows to show the way water moves through the water cycle.

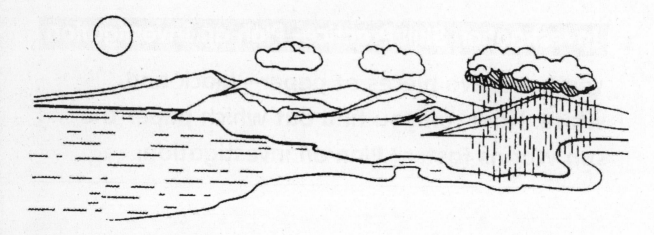

4. The day is very sunny. Why do you think the street is hotter than on a cloudy day?

_____

_____

_____

_____

VOCABULARY GAMES
and CARDS

Contents

Vocabulary Games

You can use the word cards on pages CS81–CS118 to play these games.

> ## Get Off My Back!
>
> **You will need** word cards, tape, paper, and pencil
>
> **Grouping** whole class, large group, or small group
>
> 1. Tape three word cards to another player's back. Do not let that player see the words. Have that player tape three cards to your back. Everyone playing the game should have three word cards taped to his or her back.
>
> 2. Ask other players to give hints about each word on your back. Guess each of the three words.
>
> 3. If you guess a word correctly, move the word from your back to the front.
>
> 4. The player that is the first to move all three words to the front is the winner.

Word Square

You will need word cards, paper, pencil, and crayons

Grouping individuals

1. Fold a sheet of paper in half. Fold it in half again to make four boxes.

2. Choose one word card. In the first box, write the word.

3. In the second box, draw a picture to show what the word means.

4. In the third box, write a sentence to tell what the word means.

5. In the fourth box, draw a picture to show what the word does NOT mean.

Quick Draw

You will need word cards, paper, pencil, and crayons

Grouping whole class, large group, or small group

1. Shuffle word cards. Place them in a stack face down. Draw one card from the stack. Do not show the word to the group.

2. Draw a picture of the word. Ask group members to guess the word.

3. If the word is still not known, give one clue about the meaning of the word. Ask group members to guess the word.

4. The person who guesses the word first gets to keep the word card. The person with the most word cards at the end of the game is the winner.

© Harcourt

1

above

2

air

3

animal

4

balance

air

A mixture of gases around Earth.

2

above

At a higher place.

1

balance

A tool you use to measure the mass of an object.

4

animal

A living thing that does not make its own food.

3

© Harcourt

bar graph

below

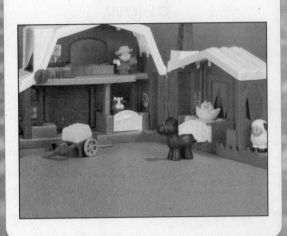

burn

carnivore

below

6

At a lower place.

bar graph

5

A graph that helps you compare numbers of things.

carnivore

8

An animal that eats only other animals.

burn

7

To change matter into ashes and smoke.

classify

communicate

compare

cool

communicate

To draw, speak, or write what you observe.

10

classify

To group things by how they are alike and different.

9

cool

To take away heat.

12

compare

To look at things to see how they are alike and different.

11

desert

13

dissolve

14

draw conclusions

15

dropper

16

dissolve 14

To mix completely with
a liquid.

desert 13

Land that gets very little rain.

dropper 16

A tool you use to place drops
of liquid.

draw conclusions 15

To use what you observe to
figure out what happened.

© Harcourt

environment

17

evaporate

18

float

19

food

20

evaporate

18

To change from a liquid into a gas.

environment

17

All the things that are in a place.

food

20

Something a living thing needs to live and grow.

float

19

To stay on top of a liquid.

food chain

forceps

forest

freeze

© Harcourt

forceps

A tool you use to separate things.

22

food chain

Shows the food living things eat.

21

freeze

To change from a liquid to a solid.

24

forest

Land that is covered with trees.

23

© Harcourt

gas

25

hand lens

26

heat

27

herbivore

28

hand lens

26

A tool you use to make small objects look bigger.

gas

25

Matter that will completely fill its container.

herbivore

28

An animal that eats only plants.

heat

27

Energy that warms.

hypothesize

infer

investigation skills

land

© Harcourt

infer

To use what you observe to tell why something happened.

30

hypothesize

To tell what you think before you test something.

29

land

The solid part of Earth's surface.

32

investigation skills

The skills people use to find out things.

31

leaves

33

liquid

34

magnifying box

35

make a model

36

34

liquid

Matter that takes the shape of its container.

33

leaves

The parts of a plant that take in light and air to make food.

36

make a model

To make an object to show how something works.

35

magnifying box

A tool that makes things you put into it look bigger.

mass

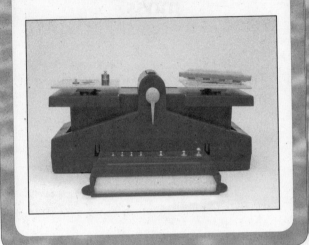

matter

measure

measuring cup

matter

38

What everything is made of.

mass

37

The measure of how much matter something has.

measuring cup

40

A tool you use to measure liquid.

measure

39

To find the size or amount of something.

melt

mix

mixture

nest (noun)

mix

42

To put together.

melt

41

To change from a solid to a liquid.

nest (noun)

44

A shelter that birds and some other animals make.

mixture

43

Something that is made up of two or more things.

nest (verb)

observation

observe

ocean

© Harcourt

observation

46

Something you learn by using your senses.

nest (verb)

45

To make a home in a shelter.

ocean

48

A large body of salt water.

observe

47

To use your senses to find out about things.

plan an investigation

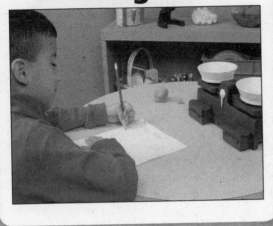

49

plant

50

pollen

51

predator

52

© Harcourt

plant

50

A living thing that makes its own food.

plan an investigation

49

To think of a way to check out an idea.

predator

52

An animal that hunts other animals for food.

pollen

51

A powder that flowers need to make seeds.

predict

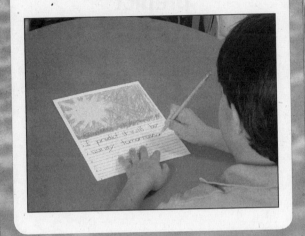

property

rain

roots

property

54

One part of what something is like.

predict

53

To use what you know to make a guess about what will happen.

roots

56

The parts of a plant that hold it in the soil and take in water and nutrients.

rain

55

Drops of fresh water that fall from clouds.

ruler

science tools

season

sequence

science tools

58

Tools that scientists use to find out things.

ruler

57

A tool to measure how long or tall an object is.

sequence

60

To put things in the order in which they happen.

season

59

A time of year.

shelter

61

sink

62

snow

63

soil

64

sink 62

To fall to the bottom of a liquid.

shelter 61

A place where a person or an animal can be safe.

soil 64

The top layer of Earth.

snow 63

Frozen water that falls from clouds.

solid

state

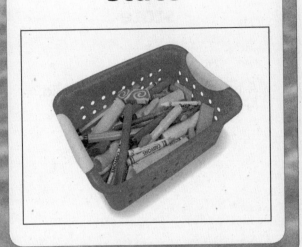

sunlight

tape measure

state
66

A form of matter—solid, liquid, or gas.

solid
65

A kind of matter that keeps its shape.

tape measure
68

A tool you use to measure around an object.

sunlight
67

Light that comes from the sun.

© Harcourt

teeth

69

tell

70

temperature

71

thermometer

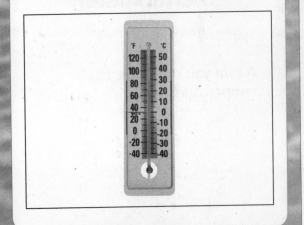

72

tell

To share what you observe.

70

teeth

Parts of the body you use to bite and chew food.

69

thermometer

A tool you use to measure temperature.

72

temperature

The measure of how hot or cold something is.

71

water

73

water cycle

74

weather

75

wind vane

76

water cycle

74

The movement of water from Earth to the air and back again.

water

73

A clear liquid found in Earth's lakes, rivers, and oceans.

wind vane

76

A tool you use to measure the direction of the wind.

weather

75

What the air outside is like.